YOUR KNOWLEDGE HAS VALUE

- We will publish your bachelor's and master's thesis, essays and papers

- Your own eBook and book - sold worldwide in all relevant shops

- Earn money with each sale

Upload your text at www.GRIN.com and publish for free

What distinguishes food insecurity in Dryland areas in Ethiopia from the highlands?

Hama Mamo Sagaro

Bibliographic information published by the German National Library:

The German National Library lists this publication in the National Bibliography; detailed bibliographic data are available on the Internet at http://dnb.dnb.de.

ISBN: 9783346791238
This book is also available as an ebook.

BULE HORA UNIVERSITY

College of Agricultural Science

Department of Rural Development & Agricultural Extension

Course Title: Graduate seminar

Title: - **What distinguishes food insecurity in Dryland areas in Ethiopia from the highlands?**

by

Hama Mamo Sagaro

December 2022

Bule Hora, Ethiopia

ACKNOWLEDGEMENT

Above all, I would like to thank the **Almighty God** for his unreserved gift and for being with me in all aspects of my life, especially in my academic performance. Besides, first and foremost, I thank my graduate seminar Course Convenor and advisor **Tinsaye Tamerat (Ph.D.)** for whom I am very appreciative and express my gratitude for his valuable advice, energetic encouragement, suggestion, insight, open-minded views, help, and guidance from the initiation to the completion of these Graduate seminar paper.

Finally, this might be a good opportunity to acknowledge all members of my family, relatives, and friends for all their encouragement and support throughout my study.

ABBREVIATION

CSA	Central Statistical Agency
DPPC	Disaster preparedness& Prevention Commission
EC	European Commission
FAD	Food Availability Decline
FAO	Food and Agriculture Organization
FDRE	Federal Democratic Republic of Ethiopia
GDP	Gross Domestic Product
MDER	Minimum Dietary Energy Requirement
PSNP	Productive Safety Net Program
RRC	Relief and Rehabilitation Commission
UNOCHA	United Nations Office for the Coordination of Humanitarian Affairs
UN	United Nation
UNDP	United Nations Development Program
US	United States
USAID	United States Agency for International Development
USD	United States Dollar
WB	World Bank
WFP	World Food Program
WFS	World Food Summit
WHO	World Health Organization

ABSTRACT

Food security and insecurity are terms used to describe whether or not households have access to sufficient quality and quantity of food. Food insecurity is a condition in which people lack the basic food intake necessary to provide them with the energy and nutrients required for fully productive lives. It can either be temporary (transitory food insecurity) or continuous (chronic food insecurity).

Like many developing countries, Ethiopian farmers in the highlands predominantly practice subsistence farming and are often subject to food insecurity. The main objective of the seminar is to review the status of household food insecurity in Ethiopia with a special focus on dryland areas. A review of the status of household food insecurity is vital because it provides information that will enable effective measures to be undertaken to prove food security status and bring the success of food security development programs.

The result of the empirical review indicated, especially in dryland areas of Ethiopia, the majority of households were food insecure. Drought risks, desert locus, the spread of corona varies, protracted impacts of past poor seasons, conflict, poor household income, cost of nutritious food, and knowledge of nutritious food factors are the major drivers of food insecurity.

International non-governmental organizations, local organizations, the private sector, and the government should continue to work together to adopt drought-risk-friendly modern technologies and design new production-oriented and commercialization policies to improve food security.

The data for this review was used from secondary data sources by conducting an intensive reading of published and unpublished journals, articles, and books. Also, this paper is going to review the Causes, indicators, coping mechanisms, and Policy options to minimize household food insecurity in dryland areas of Ethiopia.

1. INTRODUCTION

1.1 Background of the seminar

The issue of ensuring food security has become the agenda of concern across the world in several contexts. Worldwide data shows that the number of people undernourished and severely food insecure people in 2018 was about 821.6 and 704.3 million people respectively. In the same year, in Africa, 256.1 and 277 million people respectively are undernourished and severely food insecure. In Ethiopia, the average number of undernourished people in three years (From 2016-2018) is about 21.6 million (FAO, 2018)

It is widely accepted that food is a basic means of sustenance and a human right. Having enough food in terms of quantity and quality for all people is an important factor for a healthy and productive life (Sani & Kemaw, 2017). Adequate intake of quality food is needed condition to be well-nourished and indicates the food security status of a household. In contrast, food insecurity exists when people lack access to an adequate and safe supply of food on a stable basis. (FAO., 2005)

Food security is a topic of keen interest to policymakers, practitioners, and academics around the world in large part because the consequences of food insecurity can affect almost every facet of society (Feleke, 2019) For example, the food price crisis and subsequent food riots in 2007–2008 highlighted the critical role of food security in maintaining political stability (Jones, 2013) Reports indicate that about 795 million people in the world were food insecure, with many more sufferings from "hidden hunger" caused by micronutrient or protein deficiencies (FAO I. &., 2015)

As a part of Africa and the developing world, Ethiopia is one of the most food-insecure and famine-affected countries as a large portion of the population is affected by food insecurity. The Ethiopian economy is mainly dependent on agriculture which is vulnerable to different shocks, seasonality, and trends (Bedemo, 2014). For example, in the country, most areas are exacerbated by low production and crop loss mainly caused by seasonal unpredictable and sporadic rainfall and climate shocks, the occurrence of drought which often results in low farm production, widespread lower income, and subsequent food shortages and famines.

According to (Bezu, 2018) in Ethiopia, nearly 33 million people are suffering from chronic undernourishment and food insecurity with the highest percentage of the food-insecure population living in dryland areas. According to the report of the ministry of agriculture and rural development (MoARD., 2009) arid and semiarid rangelands of Ethiopia comprise nearly 13% of the population, while these areas constitute about 63% of the country's landmass.

According to the report (FAO, 2018), the main causes of food insecurity in most drylands of Ethiopia are prolonged drought, conflict, political instability, crop disease, flooding, protracted impacts of past poor seasons, desert locusts, poor household income, and cost of nutritious foods and knowledge on nutritious food factors. And also, the report shows, in Ethiopia, prolonged drought conditions are severely affecting the livelihoods in most southern and southeastern dryland areas of south nation nationality people which is a part of the state (region), southern Oromia, and southeastern Somali Regions, where cumulative seasonal rainfall was up to 60% below average

To reverse this problem, the government of Ethiopia has been formulating and implementing various strategies and programs like a productive safety net program (a platform to provide emergency-related support to vulnerable households in need of relief cash/food in times of shocks) and agricultural development - led industrialization (example, Growth and Transformation Plan one and two) in which food security strategy is a key component. To foster broad-based development sustainably, the growth and transformation Plan (GTP I) (2010/11- 2014/15) was implemented. Generally, the aim of the plan has significantly increased the share of industry in the economy along with the rise in agricultural production. Hence, according to Dube et al. (2019) during the GTP I implementation period, there was a positive achievement in the economic growth of the country such as a double-digit growth rate of real GDP and a decline in the incidence of poverty from 38.7% headcount poverty index in 2005 to 29.6% in 2010/11. Similarly, GTP II (2015/16- 2019/20) is a continuation of GTP 1 and gives more emphasis on humanitarian challenges arising from climatic change that continued affecting the economic growth of the country, production of high-value crops and livestock, and market orientation. (Dube, 2019)

In addition, pastoral and agro-pastoral households in the dryland areas of Ethiopia use different kinds of coping mechanisms to reduce the incidence of food insecurity. As coping strategies, the mobility of pastoralists exploiting the animal feed resources along different ecological zones represents a flexible response to a dry and increasingly variable environment because animal byproduct like milk is the main source of food. This means pastoralists are movable in searching for better animal feed sources because in most dryland areas the rainfall is very erratic. Hence, pastoralists move to areas having rainfall and feed their animal grass which is the only source of animal feed. However, constraints on pastoral mobility, such as changes in land use, tenure regulations, and borders, have undermined the whole pastoral system result adults in the problem of food insecurity continuing to persist in the country (FAO, 2018)

2. STATEMENT OF THE PROBLEM OF THE SEMINAR

Though Ethiopia has abundant natural resources, most of its socioeconomic indicators are extremely low. In Ethiopia, food shortage has aggravated the already poor economy of the country. Both chronic and transitory problems of food insecurity are widespread and severe in both rural and urban areas of the country (FDRE, 2002) However, a lot of studies conducted so far in the field give more emphasis to the rural area of the country (Eden & Nigatu, 2009) But such partial assessments do not verify situations at the grass root level and hide the true food insecurity problem of the country. Furthermore, such studies or reviews do not look at the underlying causes of food insecurity in households in Ethiopia. The extent of the food insecurity problem differs from place to place and by social position and actual living conditions.

The country is also known to possess the largest livestock population in Africa. However, poverty and food insecurity remain the major challenges to achieving economic development in Ethiopia and especially in the rural area of the country. This is due to the subsistence nature of Ethiopian agriculture, its mere dependence on rainfall, and the existing backward technology, which has made peasants highly vulnerable to famine and food insecurity.

Since the problem is more severe, especially in the dryland areas of the country, the government of Ethiopia has recently appealed to its international partners for emergency food assistance to feed 10.2 million people and for special nutritional programmers" for more than 2.1 million, including 400,000 severely malnourished children. In addition, over 8 million vulnerable and food-insecure people receive support under the Productive Safety Net Programme (Nkunzimana, 2016)

Food insecurity is a real and major problem in dryland areas. Despite this, reviews on household-level determinants of food insecurity and local coping strategies are rare in the area. Hence, this review was initiated to address the knowledge gap as far as food insecurity and local coping strategies in the pastoral and agro-pastoral contexts are concerned. Thus, this review synthesizes the status of household food insecurity and drivers of food insecurity in dryland areas which provides important information to policymakers, practitioners, academics, and other concerned bodies to intervene and improve the food security status of the households in the country.

3. OBJECTIVE OF THE SEMINAR

3.1. General Objective of the Seminar

➤ To review the status of Household Food Insecurity in dryland areas of Ethiopia

3.2. Specific Objectives of the Seminar

➤ To review the causes of Household Food Insecurity in Ethiopia
➤ To review indicators of Household Food Insecurity in Ethiopia
➤ To review Coping mechanisms and Policy Options to minimize Household Food Insecurity in Ethiopia

4. RESEARCH QUESTION

✓ What is the cause of household food insecurity in dryland areas of Ethiopia
✓ What are the indicators of household food insecurity in dryland areas of Ethiopia
✓ What are the Coping mechanisms and Policy Options to minimize food insecurity in dryland areas of Ethiopia

5. SIGNIFICANCE OF THE SEMINAR

A review of the status of household food insecurity in dryland areas of Ethiopia is vital because it provides information that will enable effective measures to be undertaken to improve food security status and bring the success of food security development programs.

It will also enable development practitioners and policymakers to have better knowledge as to where and how to intervene in rural areas to bring food security or minimize the severity of food insecurity. The positive role of the agricultural sector in terms of ensuring national food security is measured by the contribution of domestic food production to national food availability and reduced dependence on food imports.

Although the food deficit in Ethiopia has never been overcome during the last three decades, domestic production is playing a significant role to reduce the gap and dependence on food imports to a limited extent. The national food security goal is to attain food self-sufficiency by increasing the use of a package of modern farm inputs through an agricultural extension program, as well as improved husbandry conditions in livestock-producing areas. Improved domestic food production in the Ethiopian context will have the following positive social benefits:

> Reduced dependence on food imports and saving foreign exchange; allocating the saved foreign exchange for other alternative public uses will increase the social benefits;
> It contributes to human resources development. This would be possible through improved nutritional status and reduced health costs in society (minimizing health conditions that arise from poor nutrition and health consequences);
> It will enable the supply of a healthy and productive labor force required for economic growth.

6. METHODOLOGY

The data for this review was used from secondary data sources by conducting an intensive reading of published and unpublished journals, articles, and books.

7. RESULTS AND DISCUSSIONS

7.1. Review of the Status of Household Food Insecurity in Dryland Areas of Ethiopia

7.1.1. Concepts and Definitions of Food Insecurity

In the mid of 1970s and 1974s world food conference was held to solve the problem of world food crises and major famines around the world. Food security and insecurity are the terms used to describe whether or not households have

access to sufficient quality and quantity of food. With progress in time and the severity of the problem, food security issues gained prominence and great attention at the global, national, household, and individual levels. Such progressive work by scientists led to redefining the scope and depth of the food security concept. For instance, (Duffour, 2010) explained the concept stating that food security at the global level does not guarantee food security at the household or individual level. Without much change in the basic concepts, different institutions and organizations define food security in different ways. According to (FAO, 2008) food security is a situation that is achieved at the individual, household, national, regional, and global levels when all people, at all times, have physical and economic access to sufficient, safe, and nutritious food that meets their dietary needs and food preferences for an active and healthy life. On the other hand, in recent studies, food security is defined as adequate availability of and access to food for households to meet the minimum energy requirements recommended for an active and healthy life (Hussein W., 2013) .

According to the 1996 World Food Summit Food security exists "when all people, at all times, have physical and economic access to sufficient safe and nutritious food that meets their dietary needs and food preferences for an active and healthy life". Based on this definition (FAO, 2008) developed **four main dimensions of food security** which are **food availability, food accessibility, food utilization, and stability**.

Examining the dimensions of food security provides a more comprehensive picture, and can also help in targeting and prioritizing food security and nutrition policies and programs.

Food availability: refers to the presence of food at global, national, household, and individual levels, example when "sufficient quantities of appropriate, necessary types of food from domestic production, commercial imports, commercial aid programs, or food stocks are consistently available to individuals or nations." Hence, food availability is largely a function of macroeconomic factors (Anderson, 2015) The food availability indicators capture not only the quantity but also the quality and diversity of food. For assessing food availability, adequacy of dietary energy supply, the share of calories derived from cereals, roots, and tubers, average protein supply, and the average value of food production should be analyzed.

Food access: refers to the resources that households have to obtain food, either through their products or through purchase. So, individuals need to have assets or incomes to produce, and purchase to obtain foods needed to maintain their consumption. Hence, food access is largely related to household income and own production (Anderson, 2015) . Food access depends on; income available to the household, the distribution of income within the household, the price of food in the market, and other factors worth mentioning are individual access to the market and social and institutional rights.

Food utilization: refers to the nutritional benefits derived from food consumption which are related to proper food processing, storage techniques, adequate knowledge of nutrition; and adequate health and sanitation services exist. Hence food utilization is largely related to nutrition, health, and sanitation (Anderson, 2015). The same to this (Jrad, 2010) defines food utilization as the 'proper biological use of food, requiring a diet that contains sufficient energy and essential nutrients as well as knowledge of food storage, processing, basic nutrition, child care, and illness management'.

Food stability: refers to the stability of all other dimensions of food security over time. Even if your food intake is adequate today, you are still considered to be food insecure if you have inadequate access to food periodically, risking a deterioration of your nutritional status.

Adverse weather conditions, political instability, or economic factors (unemployment, rising food prices) may have an impact on your food security status (FAO, 2008). Therefore, for food security to be insured at global, regional, national, household, and individual level food stability should be maintained.

Food insecurity, on the other hand, is a situation that exists when people lack secure access to sufficient amounts of safe and nutritious food required for normal growth and development and active and healthy life (WFP, 2009) It is a dynamic phenomenon: its impact varies depending on its duration, its severity, and the local socioeconomic and environmental conditions.

7.1.2. Types of food insecurity

Food insecurity can be transitory (when it occurs in times of crisis), seasonal or chronic (when it occurs continuingly). A person can be vulnerable to hunger even if he or she is not hungry at a given point in time (Zezza, 2003.)

Chronic food insecurity is a long-term or persistent condition that occurs when people are unable to meet their minimum food requirements over a sustained period. It results from extended periods of poverty, lack of assets, and inadequate access to productive or financial resources (FAO, 2008). Contrarily, **transitory food insecurity** is short-term and temporary and occurs when there is a sudden drop in the ability to produce or access enough food to maintain a good nutritional status. This means transitory food insecurity is caused by short-term shocks and fluctuations in food availability and food access, including year-to-year variations in domestic food production, food prices, and household incomes (FAO, 2008). Both chronic and transitory problems of food insecurity are widespread and several in Ethiopia.

7.1.3. Food insecurity situations in Ethiopia

In Ethiopia, food insecurity is highly prevalent in the moisture-deficit highlands and the lowland pastoral and agro-pastoral dryland areas. Abule et al. (2005) identified the six regions of dryland areas in Ethiopia, considering the environmental conditions, floristic composition, and productivity values (Figure 1). These arid and semiarid rangelands of Ethiopia as shown in Figure 1, comprise nearly 13% of the population, having 63% of the country's landmass (MoARD., 2009)

Even in years of adequate rainfall and good harvest, the people, particularly in dryland pastoral and agro-pastoral areas remain food insecure and in need of food assistance. In dry (lowland) areas, droughts have become frequent and more severe in recent years and are one of the most important triggers of malnutrition and food insecurity in the country (Dominguez, 2010). In most of these areas, the rainfall is raining in a very short period, starting from June to August; the remaining time of the year is dry. Even this time, the level of rainfall is very erratic resulting in droughts and other related disasters (such as crop failure, water shortage, and livestock disease, land degradation, limited aid household assets, low income (Mohamed, 2017)

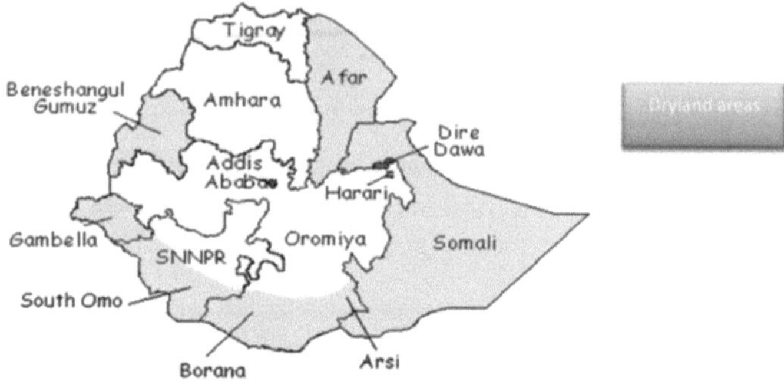

Figure 1. Location map of the major dryland areas in Ethiopia. (Source: Abule et al. (2005).

According to (FSIN., 2020) among intergovernmental authority for development (IGAD) countries, Ethiopia was the most food-insecure country in the region with 8.1 million food-insecure people in need of urgent action, followed by Sudan with 6.2 million, and South Sudan with 6.1 million. The problem is more severe in the dryland areas of the region because, mostly, prolonged dry conditions and flash floods negatively affected pastoral and agro-pastoral livelihoods by causing below-average crop production, and pasture, as well as limiting water sources for both, and resulting in chronic and acute food insecurity is prevalent. As a result, these regions have become heavily dependent on external food aid.

The Integrated Food Security Phase Classification [IPC], 2020 classifies food insecurity into five phases; phase one (People minimally food insecure), phase two (People under Stress), phase three (people in crisis), phase four (people in an emergency) and phase five (people in catastrophe). The report also evidenced that of the 8 million food-insecure people in Ethiopia, 21% are severely food insecure at the IPC Phase three-level, 38% at IPC phase two, 34% at IPC phase one, 6-% IPC phase four, and 0-% IPC phase five. The analysis includes all food-insecure households irrespective of whether they benefit from a productive safety net program (PSNP), including current internally displaced persons or returnees

7.2 Factors that cause food insecurity

According to the report of the (WFP F. N., 2021) In Ethiopia, causes of food insecurity have always been complex given that multiple factors affect food security. The factors that cause food insecurity are wide and vary from place to place but in this literature, the following major causes will be discussed.

Food availability

- Domestic production
- Import capacity
- Food stocks
- Food aid

Stability of supply and access

- Weather variability
- Price fluctuations
- Political factors
- Economic factors

Access to food

- Purchasing power
- Income of population
- Transport and market structure

Food utilization

- Food safety
- Hygiene and manufacturing practices
- Diet quality and diversity

Figure 2. Determinants of each pillar.

Source: Al et al. (2008), and Marion (2011).

7.2.1 Population Pressure

Home to about 115 million people, Ethiopia is the second-most-populous country in Africa and has one of the region's fastest-growing economies (WorldBank, 2021). Food insecurity is a major and ever-worsening problem. Rapidly increasing population pressure is one of the underlying causes of food insecurity (USAID, 2012.)

Population pressure coupled with droughts and other unfavorable weather conditions is a challenge to famine prevention in Ethiopia (Alexander, 2009.). Most of the countries with the highest numbers of people facing food insecurity also have high fertility rates and rapid n   growth. This increases the challenge of adequately meeting nutritional needs. Sub-Saharan Africa has the highest population growth rate in the world (United Nations Population Division, 2009.).

A large population reduces income per head, expenditure per head, and capita food consumption. The likely explanation is that in an area where households depend on less productive agricultural land, increasing population results in increased demand for food. This demand, however, cannot be matched with the existing food supply from own production and this ultimately ends up with the household becoming food insecure (Robert, 2013) .

7.2.2 Natural factor

Natural disasters such as rainfall, drought, climate change, and flooding are said to reduce food production for a particular period. In the case of Ethiopia, there is no doubt that droughts have created severe food shortages. In most dryland areas of Ethiopia, the level of rainfall is very erratic and rains in a short period (mostly, June to August) flooding is also a major problem at this time. In addition, there are increased daily air temperatures throughout the year. These climate changes (weather variability) affect cropping and pasture which reduces crop yield and results in poor dietary diversity and consumption. Hence, erratic rains, uneven distribution, and long dry spells resulted in late planting and poor crop development, poor pasture regeneration, and low water availability are the key drivers of food insecurity that negatively impact the four pillars of food security in Ethiopia (Figure 2). Similarly, COVID-19 has affected the transport and market structure and resulted in economic disruptions. In addition, in the future, it may be one major source of food insecurity that negatively impacts the four pillars of food security in Ethiopia depending on its spread level throughout the country.

The report also indicates, currently, the desert locust is a problem in some areas of Ethiopia which affect crop production and pasture. It spread over most Dryland areas of Ethiopia, across areas of Somali, Afar, eastern lowlands of Amhara, Oromia, and south nation nationalities and people regions because desert locusts would invade a larger extent of lowland and main cropping areas due to high temperatures. Hence, it is one factor that results in food security which directly and indirectly affects the four pillars of food insecurity (Figure 2).

Markets are the major sources of food. However, in most dryland areas of Ethiopia, market prices are not favorable for households, following a persistent increase in cereal prices due to shortages coupled with other factors. The recent spike in food prices, largely driven by a limited supply of cereals and erupting conflicts in most parts of the region, affected households dependent on staple foods. The conflict facing is recurrent inter-communal, especially over water and pasture. The conflicts have limited the movements of households and their livestock and led to disruptions in the major markets, limiting physical access to food and affecting prices, and continue to affect food security (IPC, 2020)

7.2.2.1. Drought

(Reported By (UNOCHA., 2021)

Ethiopia is experiencing a prolonged drought after three consecutive failed rainy seasons since late 2020 affecting 6.8 million people living in Oromia, SNNP, Southwest, and Somali: several areas in southern and southeastern Ethiopia, including in the regions of Somali (10 zones), Oromia (8 zones), Southwest (1 zone) and Southern Nations, Nationalities, and Peoples - SNNP (7 zones). People living in these same areas have barely managed to recuperate from the severe drought in 2017 to witness again such harsh conditions, the first signs of which started appearing towards the end of 2020. The conditions continued to worsen with the successive failed seasons in 2021. The next season in March/April 2022 might also be well below normal, making it the fourth consecutive failed rainy season thus leading more people into an alarming situation.

The drought is compromising fragile livelihoods heavily reliant on livestock and causing worsening food security and nutrition while eroding coping strategies for the most vulnerable: The number of livestock dying from lack of food and water is staggering and increasing by the day. The high number of livestock deaths is an important indicator of this alarming situation. The numbers reported by January 2022 have by far exceeded the estimates from 2021. While 68,000 livestock deaths had been estimated in November, revised to 172,000 in December, the figures from January stand at more than 260,000 in January across Somali, southern Oromia, and SNNP regions. An additional 2 million livestock are at risk across affected areas. Overall, the drought is currently affecting the livelihood of some 6.8 million people across the drought-affected areas.

The drought is worsening food insecurity and malnutrition, while pastoralists are forced to walk and travel long distances in search of water and pasture:

Drought is the major cause of the loss of food production in areas where agricultural activity has been overwhelmingly dependent on rainfall. In cases like Ethiopia, the contribution of irrigated land in agriculture is very minimal and the whole food production activity has depended on rain-fed agriculture; the disruption of rain for a season has brought a massive food shortage and consequently hunger and famine. Thus, it is clear that drought and rainfall variability constrain the food security status the rural agricultural-dependent farmers.

Table .1 Chronology of drought-related food security crises in Ethiopia

Year of event	Major relative incidences
1953	The food security crisis in Wollo and Tigray. Raya Azebo is the most vulnerable from Tigray.
1957-58	The food security crisis in Tigray, Wollo, and south-central Shewa. About 1 million farmers in Tigray might have been affected, with about 100, 000 being displaced.
1962-66	Many parts of northeastern Ethiopia suffered from droughts and Food security crises, and Tigray and Wollo were severely hit.
1973-74	This was one of the most significant food security crises which affected parts of eastern Harare, SNNPR, and the Bale lowlands. About100, 000 to 200,000 people died as a result of this extensive crisis.
1977-78	Most parts of the Wollo were severely hit by the food security crisis owing to erratic rainfall, pest damage, and frost actions. About 500,000 farmers were affected.

1984-85	This was the most serious one affecting over eight million people and causing the death of one million Ethiopians. Most parts of Ethiopia including relatively food-secure areas like Wolayta, Kambata, and, Hadiya were affected by severe food insecurity. Drought and crop diseases were the main drivers of the food security crisis in this case.
1987-88	Tigray, Wollo, and Gonder were severely affected due to drought and civil wars.
1990-92	Rain failure and regional conflicts resulted in approximately 4,000,000 people being affected.
1993-94	Widespread food insecurity, but few deaths or cases of displacement were reported because of early responses by the government and international aid organizations.
1999-2000	Three years of successive poor rains in the Somali region led to 100,000 deaths of citizens.
2003-04	Over 13 million people were affected, but the response mitigated the worst potential outcomes.
2008-09	Almost 3 million people were affected which majority from pastoral areas of the country.

| 2011-13 | A severe food security crisis occurred in the southeastern lowlands. In the pastoral areas of Afar, Somali, and Borana (Oromia) the quality and quantity of livestock decreased. |
| 2015-16 | Consecutive failure of two rainy seasons has had a profound impact on the lives and livelihoods of millions. Due to the o El Niño drought, more than 27 million people become food insecure and a total population of 18.1 million people required food assistance in 2016, and a total of 40 percent of the Ethiopian population have been affected. This was the strongest drought that has been faced in Ethiopian history. But the number of deaths due to this shock was not reported |

Source: [Catley et al., (2016), DFID (2014), Webb et al., (1992)]

From the above table, we can conclude that every two- or three years Ethiopia faces one drought year that leads to huge food crises and the loss of millions of livestock numbers but also human death in different drought-prone parts of the country. 1984/85 was a serious drought as reported by different agencies like Relief & Rehabilitation Commission (1985). However, 2016 is the strongest drought in the history of the country which indicates vulnerability to drought in Ethiopia increasing from time to time

7.2.3 Economic factors

Many studies conducted in different parts of Ethiopia showed that farmland, credit, livestock holding, and access to different productive assets are affecting the food security status of rural households in Ethiopia.

More land size holding means more cultivation and more possibility of production which in turn increases farm income and improves food security (TesfayeB, 2003.). According to Adugna E (2008) conducted his study in the Boloso Sore district wolayita zone, shortage of oxen, lack of farm input, and land shortage are the most influential causes of food insecurity.

Similarly, according to (Amsalu Mitiku, 2012.) study done shashemene district Oromia region logit model result showed that total cultivated land, total annual farm income per adult equivalent, total off-farm income, and livestock size have a positive and significant relationship with food security.

7.2.4 Socio-cultural factors

In our country, Ethiopia gender division of labor due to cultural factors constrain productivity and food security. According to the study conducted by (McBriartyk., 2011), in all his study areas of rural Ethiopia, it is not socially acceptable for women to plow. This is a major constraint to preparing fields for planting on time for the season. Single women, therefore, had to resort to begging neighbors and waiting until everyone else has finished their plowing, otherwise, they must pay someone to do it for them.

(TsegayeG, 2009) reviewed those Socio-cultural events such as eating habits and food preferences, cultural ceremonies, and festivals also influence the food security status of the given communities and way of saving or expenditure, also directly or indirectly affects the food security situation of that particular community.

7.2.5 Institutional factors

Several studies agree that poor infrastructure including roads; schools and health services constrain productivity and thereby bring food insecurity and dependence on food aid.

One study conducted on factors contributing to rural women's food insecurity in Ethiopia showed that the lack of roads and transport results in fewer market opportunities, less influence from outside, and added difficulty in acquiring medical treatment or access to information and it seriously affects their attainment of food security (McBriartyk., 2011).

7.2.6. **Conflict/insecurity**

Conflict is a leading cause of food insecurity and hunger in several parts of the world, undermining food security in multiple ways and creating access problems for governments and humanitarian agencies who often struggle to reach those most in need. According to (FSIN., 2020), conflict and insecurity are the major drivers of food insecurity in eighteen countries, and the number of food-insecure people across the world has been increasing over time.

Tigray, Afar, Amhara, Benishangul-Gumuz, and Oromia experienced high levels of violence, displacement, and destruction of livelihoods in 2021 (FEWSNET, 2021.). In Tigray and neighboring Amhara and Afar regions, conflict had a dramatic impact on food security, mainly through large-scale displacements, and movement limitations that impaired livelihood activities, market functioning, and access to basic services and humanitarian assistance. Households faced losses of income from agricultural, casual, and salaried labor, with salaries not paid to most public and private sector workers (FAO-GIEWS., 2021)

In Southern Tigray, insecurity reduced the areas planted for the secondary 2021 season of Belg crops. This, coupled with delayed and erratic rainfall, led to a near failure of the harvest in July, while sowing operations of the major 2021 Meher crops in May–June were also affected by insecurity and lack of inputs, resulting in a substantial reduction of the planted area. Although average to above-average June–September Kiremt rains had a positive impact on yields, crop production was estimated at 60 percent below the already poor 2020 main harvest, resulting in the third consecutive season with reduced production since the start of hostilities in November 2020 (FAO-GIEWS., 2021).

Since mid-2021, humanitarian access to Tigray has been heavily constrained by armed clashes in boundary areas with Eritrea, Amhara, and Afar regions (UNOCHA., 2021). The region-wide shutdown of banking and communication services and lack of fuel due to conflict impeded the delivery of food assistance within Tigray, forcing humanitarian partners to halt or significantly reduce operations (UNOCHA., 2021).

7.3. Indicators of food insecurity

Assessment of food insecurity is challenging work for researchers dealing with it as there are no universally established indicators that serve as a measuring tool. It requires a multi-dimensional consideration since it is influenced by interrelated socio-economic, environmental, and political factors.

The full range of food insecurity and hunger cannot be captured by any single indicator. Instead, a household"s level of food insecurity or hunger must be determined by obtaining information on a variety of specific conditions, experiences, and behaviors that serve as indicators of the varying degrees of severity of the condition (Bickel, 2000)

The most frequently used measures include consumption and expenditure, nutritional status, coping strategies, and resource-oriented correlates.

Consumption indicators: Consumption is a better indicator of lifetime welfare than income. Income typically rises and then falls in the course of one's lifetime, in addition to fluctuating somewhat from year to year, whereas consumption remains relatively stable. Consumption, rather than income, is that households may be more able, or willing, to recall what they have consumed and spent rather than what they earned. In general, measuring food security based on consumption Shows the current actual standard of living; smooths out irregularities, and so reflects long-term average well-being; and is Less understated than income, because consumption expenditure is easier to recall (WorldBank, 2010).

Monetary expenditure indicators: Monetary poverty is a somewhat more indirect indicator of people's economic access to food, given the additional necessity of purchasing important nonfood items. Indeed, potential substitution between the demand for food and nonfood items is an important rationale for viewing poverty indicators as preferable to food-or calorie-based indicators. As per the preceding discussion, higher food prices might not reduce calorie consumption but could significantly reduce nonfood expenditures, thereby raising poverty. For this reason, many economists still advocate monetary poverty as an attractive indicator of food insecurity. Moreover, absolute poverty lines are usually linked to minimum calorie consumption requirements, providing a potentially important empirical link to food insecurity. In practice, however, poverty lines often become delinked over time (Deaton, 2009.).

Dietary diversity indicators: Dietary diversity is defined as the number of individual food items or food groups consumed over a given period. The food groups include grains, pulses, vegetables, fruit, meat/fish, milk/dairy, sugar, and oil/fat. Higher scores denote a more varied diet and are suggestive of a higher quality diet with a potential for higher micronutrient intake. Indicators of dietary diversity have been validated against dietary quality (specifically, nutrient adequacy) in both developed and developing countries (Ruel, 2003).

It can be measured at the household or individual level through the use of a questionnaire. Most often it is measured by counting the number of food groups rather than items consumed. The type and number of food groups included in the questionnaire and subsequent analysis may vary, depending on the intended purpose and level of measurement. At the household level, dietary diversity is usually considered a measure of access to food, (e.g. of households" capacity to access costly food groups), while at the individual level it reflects dietary quality, mainly micronutrient adequacy of the diet. The reference period can vary but is most often the previous day or week (WFP, 2009).

Food Security trends in Ethiopia: Ethiopia is highly prone to recurrent natural hazards. Across the country, an estimated 7.6 million (or 11 percent of the rural population) are considered chronically food insecure, meaning each year they are relying on resource transfers to meet their minimum food requirements. In addition, over the past four years between 2.2 and 6.4 million additional people were food-insecure or not able to meet their food needs in the short term due to transitional factors. They are temporarily dependent on relief food assistance. The main shock in most parts of the country is the lack of/erratic rainfall. Other shocks include localized floods, crop and livestock diseases, crop pests, and frost/hail.

The major challenge in the use of these response indicators is the difficulty associated with identifying the normal phenomenon of the household and the response of the household forced by food stress or its response to avoid risks of food stress. On the other hand, indicators related to food availability for consumption, level, and change in food intake serve as a measurement to define household food insecurity (FAO/WFP, 2012.). **Fluctuation** in the level of food production and possession of productive capital are also useful indicators.

7.4. Coping mechanisms of food insecurity

The coping strategy is defined as a mechanism by which households or community members meet their relief and recovery needs and adjust to future disaster-related risks by themselves without outside support (Tesfaye kumbi, 2005.)

Households adopt and develop diversified coping strategies and sequential responses which people used at times of decline in food availability. Different studies have identified food insecurity coping mechanisms in Ethiopia some of which are reviewed in the following manner.

To cope with these problems Ethiopian people in dryland areas, use the sale of livestock, agricultural employment, and migration to other areas, requesting grain loans, sales of wood or charcoal, small-scale trading, and limiting the size and frequency of meals as major coping mechanisms (Abduselam, 2017).

 According to (Abdirahman, 2015) sale of more livestock than usual, borrowing of food, reducing the number of a meal, reducing the size of meals, sale of firewood, and charcoal, seasonal migration, seeking alternative or additional jobs, relying on less preferred and less expensive food, seeking relief assistance, becoming temporary trade, household splitting, consume wild food, remittance, participating in cash basis project works were the common coping mechanisms practiced by households ho faced food insecurity problem.

According to (Birara E, 2015).) The sale of wood or charcoal, small-scale trading, income transfer systems, limiting the size and frequency of meals, sale of livestock, agricultural employment, and migration were major coping strategies.

The rural dwellers of Ethiopia used different coping strategies to cope with the existing food insecurity including a reduction of the number and quantity of meals per day; diversification of livelihood incomes, wages, and migration. In addition to the coping mechanisms used by rural households, the government of Ethiopia used different strategies to mitigate food insecurity in Ethiopia including food aid, implementation of a productive safety net program, and other food security reduction programs (YenesewS, 2015)

7.5. Policy options to minimize food insecurity

Food aid, today, is mainly considered an instrument in addressing both transitory and chronic types of food insecurity in low-income countries. It is noted that the humanitarian agencies, or donors, implement food aid programs in these countries to give immediate response to the needy people, increase income sustainability, improve agricultural productivity, and improvement in health and nutrition among the residents. Moreover, it leads to improvement in the availability of food supplies at the national or regional level, or increases access to food at household levels through higher home production of food crops, market purchase, and/or other means or to make more effective utilization of food at the individual level to meet human biological needs (USAID, 1999.)

According to the African development bank food security brief (2011), creating Policies and regulations that are conducive to enhancing regional trade, sustainable access, and use of natural resources and private investments are also essential to the success and sustainability of the benefits derived from any intervention and thereby bringing food security.

According to Mukherjee (2008), the following strategies are appropriate to eliminate hunger

1. Strengthen productivity and incomes: Diversification and growth of the economy; low-cost, simple technology (water management, use of green manures, crop rotation, and agroforestry); rural infrastructure development (roads, electricity, etc.); provision for improved irrigation and soil nutrition, natural resource management (including forestry and fisheries); market and private sector development, Food safety and quality Agricultural research, extension and training.

2. Linkages maximizing synergy: Democratic Governance Vibrant Civil Society Strong, Local food procurement for safety nets, Support to rural organizations; Primary health care and reproductive health services; Prevention and treatment of HIV/AIDS; Asset redistribution (including land reforms); Education, especially for girls and women Potable drinking water

3. Provide direct access to food: Mother and infant feeding; Supplementary nutrition to children (such as mid-day meals in Schools) and pregnant women; Unemployment and pension benefits Food-for-work and food-for-education; Targeted conditional cash transfers Food banks and Food Distribution System for the indigent people (Safety Nets); and Emergency ratio.

8. CONCLUSION

This work has thoroughly reviewed the assessment food insecurity status of households,' focused on dryland areas of Ethiopia. In drought-prone areas of Ethiopia, drought, recurring food shortages, and famine are great challenges faced by people. A high incidence of poverty and an alarming degree of food insecurity are the features of households in dryland areas. The main determinants of poverty mainly in the dryland area of Ethiopia are limited resources, low incomes, a low level of human and social capital, as well as limited access to markets and service institutions, including those for credit, extension, and plant protection. (Ogato, 2009)

Ethiopia has experienced long periods of food insecurity. Among Sub-Saharan countries, Ethiopia is the worst of all regions in the prevalence of undernourishment and food insecurity. A large portion of the country's population has been affected by chronic and transitory food insecurity. The severity is more in dryland areas of the region where chronic and acute food insecurity is prevalent. Food-insecure households in these dryland regions ranged from 5% to 44% of the households are due to internally displaced persons in conflict, *drought risks, desert locus, the spread of corona varies*, protracted impacts of past poor seasons, disease outbreaks, poor household income, cost of nutritious food and knowledge on nutritious. As a result, these regions have become heavily dependent on external food aid. This figure also will likely increase due to the spread of the novel coronavirus (COVID-19) as travel restrictions and economic disruptions hinder peoples' livelihoods and access to food.

The households and productive-aged members of the household should participate in different income-generating activities and diversify the livelihood strategies that help them to escape from a wider state of food insecurity and undernourishment. The government could reduce food insecurity by improving groundwater use and helping to develop irrigated agricultural systems. In addition, international non-governmental organizations, local organizations, the private sector, and the government should continue to work together on providing better seeds, conservation of the natural resource base for food production, improvement in research and extension, upgrading rural infrastructure, improving market access, and special provision for people in particular need to reduce food insecurity.

BIBLIOGRAPHY

Abdirahman. (2015). Food Insecurity and Coping Strategies of Agro-Pastoral Households in Babile District of Somali Regional State, Ethiopia. Msc Thesis, Haramaya University, Haramaya.

Abduselam. (2017). .Food Security Situation in Ethiopia: A Review Study. International Journal of Health Economics and Policy, 2(3): pp. 86-96.

Alexander, A. V. (2009.). Understanding Famine In Ethiopia: Poverty, Politics And Human Rights. In: Proceedings Of The 16th International Conference Of Ethiopian Studies, Ed. By Svein Ege, Haraldaspen, Birhanu Teferra And Shiferaw Bekele, Trondheim 2009.

Amsalu Mitiku, B. F. (2012.). Empirical analysis of the determinants of rural households' food security in Southern Ethiopia: The case of Shashemene District. Basic Research Journal of Agricultural Science and Review Vol. 1(6), pp. 1.

Anderson. (2015). "USAID Office of Food for Peace Food Security Country Framework for Ethiopia FY 2016 – FY 2020". Washington, D.C.

Bedemo, A. G. (2014). The role of the rural labor market in reducing poverty in West Ethiopia. ,. Journal of Development and Agricultural Economics, 6(7), 299–308. https://doi.org/10.5897/JDAE2013.0518 .

Bezu, D. C. (2018). A review of factors affecting food security situation of Ethiopia: From the perspectives of FAD, economic and political economy theories.

Bickel, G. M. (2000). Guide to Measuring Household Food Security, Revised 2000. U.S. Department of Agriculture, Food and Nutrition Service, Alexandria VA. March 2000. P8.

Birara E, M. M. (2015).). Assessment of Food Security Situation in Ethiopia: A Review. ,. Asian Journal of Agricultural Research 9, 55-68.

Deaton, A. a. (2009.). "Nutrition in India: Facts and Interpretations." Economic and Political Weekly Vol. 44 (7):. 42–65.

Dominguez, A. (2010). Why was there still malnutrition in Ethiopia in 2008? Causes and humanitarian accountability. http://sites.tufts.edu/jha/archives/640. Journal of Humanitarian Affairs.

Dube, A. K. (2019). Agricultural development led industrialization in Ethiopia: Structural break analysis. International Journal of Agriculture Forestry and Life Sciences, 3(1), 193–201.

Duffour, K. (2010). "The Budget statement and economic policy of the government of Ghana for the 2011 financial year". Presented to Parliament on Wednesday, 18th November 2009, Accra, Republic of Ghana, pp: 1-52.

Eden, M., & Nigatu, R. a. (2009). The Levels, Determinants and Coping Mechanisms of Food Insecure Households in Southern Ethiopia Case study of Sidama, Wolaita and Guraghe Zones.

FAO. (2008). Climate change and food security": a framework document, Rome, Italy.: Food and Agriculture Organization,., AO. (2018). Global Early Warning – Early Action Report on Food Security and Agriculture. Retrieved from https://reliefweb.int/report/world/global-early-warning.

FAO, I. &. (2015). The state of food insecurity in the world, 46. WFP.

FAO. (2005). Assessment of the world food security situation. Committee on World Food Security, Thirty-First Session, May 23–26, 2005. Food and Agricultural Organization.

FAO/WFP. (2012.). FAO/WFP crop and food security assessment mission to Ethiopia. Special report. P 25-30.Available online at www.fao.org accessed march 2013.

FAO-GIEWS. (2021). . Update 11 November 2021 The Federal Democratic Republic of Ethiopia Dire food insecurity situation in northern areas due to conflict https://www.fao.org/3/cb7597en/cb7597en.pdf.

FDRE. (2002). Food Security Strategy. Addis Ababa,

Feleke, A. (2019). Determinants of household food security in Doyogena woreda, Kambata Tembaro Zone, South Nation Nationalities and People Regional State, Ethiopia. . Pacific International Journal, 2(1), 13–25.

FEWSNET. (2021.). Expanding drought and conflict are expected to drive severe food insecurity in 2022, and November 2021 [online]. [Accessed 28 February 2022]. https://fews.net/east-africa/ethiopia/key-message-update/november-2021.

FSIN., F. S. (2020). Global report on food crises. Retrieved march, 2020, from https://www.sadc.int/files/8415/8818/9448/GRFC_ 2020_ONLINE.pdf.

Hussein W., a. J. (2013). Determinants of Rural Household Food Security in Jigjiga District of Ethiopia, Kasetsart J. (Soc. Sci) 34: 171 – 180. Jigjiga.

IPC. (2020). Integrated Food Security Phase Classification. Acute food insecurity analysis. World food program (WFP). Retrieved June 2020,

Jones, A. D. (2013). What are we assessing when we measure food security? A compendium and review of current metrics. Advances in Nutrition,., 4(5), 481–505. Retrieved from https://doi.org/10.3945/an.113.004119

Jrad, S. B. (2010). ("Food security models". Policy Brief No. 33, Ministry of Agriculture and Agrarian Reform, National Agricultural Policy Center, Syrian Arabic Republic, August 2010, pp: 1-32.

McBriartyk. (2011). Factors Contributing toward Chronic Food Insecurity among Women in Rural Ethiopia. Report output of an internship with CARE Ethiopia"s Chronically Food Insecure Rural Women Program Design Team as part of a master's degree at Northumbri.

MoARD. (2009, Retrieved June, 020,). Ethiopian food security program (2010–2014). . Retrieved from http://extwprlegs1. fao.org/docs/pdf/eth144896.pdf

Mohamed, A. A. (2017). The food security situation in Ethiopia: A review study. . International Journal of Health Economics and Policy, , , 86–96.

Nkunzimana, T. C. (2016). Global analysis of food and nutrition security situation in food crisis hotspots; EUR 27879. Retrieved from https://doi.org/10.2788/669159

Ogato, G. S. (2009). (Improving access to productive resources and agricultural services through gender empowerment: A case study of three rural communities in Ambo District, Ethiopia), . . Journal of Human Ecology, 27(2, 85–100.

Robert, A. J. (2013). Determinants Of Household Food Security In The Sekyere-Afram Plains District Of Ghana.

Ruel, M. (2003). Operationalizing dietary diversity: A review of measurement issues and research priorities…ou journalf Nutrition 133:3911S-3926S.

Sani & Kemaw, B. (2017). Assessment of households' food security situation in Ethiopia: An empirical synthesis. Developing Country Studies,.,(12), 30–37.

Sani, S. &. (2017). Assessment of households' food security situation in Ethiopia: An empirical synthesis. Developing Country Studies, 7(12), 30–37.

Sani, S. &. (2017). Assessment of households' food security situation in Ethiopia: An empirical synthesis. Developing Country Studies, 7(12), 30–37.

Tesfaye kumbi. (2005.). Household food insecurity in Dodota sires district, Arsi zone. Coping strategies and policy options. A thesis submitted to the school of graduate studies of haramaya university, Ethiopia 142p. .esfayeB. (2003.). Influence of land size on household food security. The Case of Deder District of Oromiya Region. A Thesis presented to the School of Graduate Studies. Unpublished. Alemaya University, Ethiopia.

TsegayeG. (2009). Determinants of food security in rural households of the Tigray region. A thesis submitted to postgraduate studies of Addis Ababa University, Addis Ababa Ethiopia.

United Nations Population Division. (2009.). World Population Prospects: The 2008 Revision. New York: UN Population Division; FAO. 2010. "Food Security Statistics: Prevalence of Undernourishment in Total Population.

UNOCHA. (2021). Ethiopia Humanitarian Needs Overview 2021. New York. file:///Users/keo/Downloads/ethiopia_2021_humanitarian_needs_overview_hno.pdf.

USAID. (1999.). Tackling Food Insecurity and Malnutrition in Ethiopia through Diversification.

USAID. (2012.). Tackling Food Insecurity and Malnutrition in Ethiopia through Diversification. Webb, P., and Von Braun, J., 1994. Famine and Food Security In Ethiopia: Lessons For Africa. London: John Wiley.

Warning, F. .. (2018). Early Action Report on Food Security and Agriculture. Retrieved from.

WFP. (2009). . Comprehensive Food Security & Vulnerability Analysis Guidelines. United Nations World Food Programme, Rome, Italy: World Food Programme (WFP).

WFP. (2009). World Food Programme). Comprehensive Food Security & Vulnerability Analysis Guidelines. United Nations World Food Programme, Rome, Italy.

WFP, F. N. (2021). Monitoring Food Security in Countries with Conflict Situations. A joint FAO/WFP update for the members of the United Nations Security Council. Issue no. 8. . Rome.

WorldBank. (2010). World Development Indicators. Available at: http://data.worldbank.org/data-catalog/world-development-indicators. (Last accessed August 2010.

WorldBank. (2021). Ethiopia Overview: Development news, research, data | World Bank (Updated in October 2021). The World Bank in Ethiopia.

YenesewS. (2015). Causes and coping mechanisms of food insecurity in rural Ethiopia, 123–133. https://doi.org/10.5251/abjna.2015.6.5.123.133. 123–133.

Zezza, S. a. (2003.). A Conceptual Framework for National Agricultural, Rural Development, and Food Security Strategies and Policies. Rome, Italy: ESA Working Paper No. 03-17.

YOUR KNOWLEDGE HAS VALUE

- We will publish your bachelor's and
  master's thesis, essays and papers

- Your own eBook and book -
  sold worldwide in all relevant shops

- Earn money with each sale

Upload your text at www.GRIN.com
and publish for free